Organic Crop Production Technology

Gowri Vijayan

All Rights Reserved. No parts of this publication may be reproduced, stored in a retrieval system, or transmitted, in any form or by any means, electronic, mechanical, photocopying, recording, or otherwise, without the prior permission of agrihortico

© 2013 AGRIHORTICO

Table of Contents

INTRODUCTION   5

    Advantages of Organic Farming 6

    Disadvantages of Organic Farming 6

ORGANIC CROP PRODUCTION PRACTICES   7

    SEED/PLANTING MATERIAL 7

    SEED/PLANTING MATERIAL TREATMENT........................... 7

    TILLAGE.. 8

    SOIL SOLARIZATION.. 9

    MULCHING 9

    IRRIGATION SYSTEMS.. 10

    INTERCROPPING AND COMPANION PLANTING 10

    GREEN MANURING... 11

    GREEN LEAF MANURING.. 12

    CROP ROTATION 13

    MANURING AND COMPOSTING.. 14

    VERMICOMPOSTING 15

    COIRPITH COMPOSTING 17

    PANCHAKAVYA 18

    DASAKAVYA... 19

    BEEJAMRUT 19

    JEEVAMRUT 20

    BIOFERTILIZERS 20

PEST AND DISEASE MANAGEMENT 24
    Trap Crops 25
    Fumigation 26
    Bio-control agents 26
    Biological pesticides 28
    Mechanical control 29

WEED MANAGEMENT 31

ORGANIC FOOD PROCESSING AND HANDLING   ***32***

ORGANIC FOOD PACKAGING AND LABELING   ***34***

ORGANCI FOOD STORAGE AND TRANSPORT   ***34***

INTRODUCTION

According to the International Federation of Organic Agriculture Movements (IFOAM), Organic agriculture is a "production system that sustains the health of soils, eco-systems and people. It relies on ecological processes, biodiversity and cycles adapted to local conditions, rather than the use of inputs with adverse effects. Organic agriculture combines tradition, innovation and science to benefit the shared environment and promote fair relationships and a good quality of life for all involved". According to FAO, "Organic is a claim on the production process rather than a claim on the product itself". National Organic Program (NOP) of United States Department of Agriculture (USDA) defines 'Organic' as "a labeling term that indicated that the food or other agricultural product has been produced through approved methods that integrate cultural, biological and mechanical practices that foster cycling of resources, promote ecological balance, and conserve biodiversity". Synthetic fertilizers, sewage sludge, irradiation, and genetic engineering may not be used.

The origin of organic farming can be traced back to 10,000 years ago, to the Neolithic period of ancient civilizations like Mesopotamia, Hwang basin etc. The very evolution of agriculture over the millennium shows its organic roots. Ancient Indian literature like Rigveda, Ramayana, Mahabharata, Kautilya Arthasashthra etc. briefly mention use of organic inputs. Organic farming is based on the principle of health, ecology, fairness and care. Organic agriculture intends to provide high quality and nutritious food, without compromising on health and well-beings of human beings. It is a holistic management system, enhancing well-being of the entire agro ecosystem, including its biodiversity, soil biological activity and biological cycle.

Advantages of Organic Farming

- ✓ Reduces production cost by 25-30%, by avoiding synthetic fertilizers and pesticides
- ✓ Retains 40% more topsoil, thereby increasing crop yield up to 5 fold within five years
- ✓ Reduction in water use, soil erosion, nutrient management with cheaper resources, makes it more profitable
- ✓ Soil fertility is maintained in the long run
- ✓ Cattle grazing on organic farmlands are found to produce more healthy milk
- ✓ Organic products do not contain any preservatives or artificial flavors
- ✓ Organic products are more healthy and nutritional, due to absences of any synthetic chemical residues

Disadvantages of Organic Farming

- ✓ Initial yields are less, is more labor-oriented and time consuming
- ✓ Organic fertilizers release slowly, hence several applications will be required to get the desired effect
- ✓ Extensive knowledge is required to produce and certify organic

ORGANIC CROP PRODUCTION PRACTICES

One of the most important and crucial steps to organic crop production, is the selection of the propagation stock. The National Program for Organic Production of a nation clearly specifies the conditions for selection of the organic stock and cultural practices to be followed. The traditional cultural practices aim to improve the soil's water retaining capacity, aeration, infiltration rate, evaporation etc. At the same time, free use of the practices lead to soil erosion and humus deterioration. The type of cultural activities planned, depend on the crop selected and the cultivation patterns planned. This is very essential for organic style of cultivation, with crops depending heavily on the natural soil capability for growth. The crop production practices followed for organic farming are briefly discussed below:

SEED/PLANTING MATERIAL

Plant materials and seeds of organic origin should be used for propagation. In case of its unavailability, chemically untreated conventional materials can be used. Organic farming strictly prohibits the use of genetically engineered seeds, pollen, transgene plants or plant materials for cultivation. Since use of synthetic plant protection chemicals is prohibited, it will necessary to select a disease free seed stock from a resistant variety.

SEED/PLANTING MATERIAL TREATMENT

Seed treatment is used to give a fighting chance for the seeds against limiting factors. Over the years various innovative seed treatment techniques have come up, namely:

- ✓ Hot water treatment- Soaking of seeds at 53^0C for 20-30 minutes
- ✓ Use of bio fertilizers- A combination of Rhizobium/ Azotobacter with PSB
- ✓ Turmeric treatment- Seeds treated to a mix of turmeric rhizome powder with cow urine
- ✓ Use of Panchakavya extract
- ✓ Application of *Trichodermaviride* (4 g/kg seed) or *Pseudomonas fluorscens* (10g/kg seed)

TILLAGE

Tillage is the first step taken after clearing of the field for cultivation. It is also among the first stages of defense taken against weed growth, soil borne pests and pathogens. The primary objective of tillage is to reduce the hardpan formation in soil. Once the soil is loosened for further process, it also exposes the soil borne pathogens to sunlight for decontamination. The same applies for any pests or weeds present during the time. Primary tillage is intended towards conservation of crop residue and added manures in the upper layers of soil. Farmers adopting row cropping, use blind cultivation-shallow tillage, avoiding the rows. Rotary hoes, wire tooth harrows are also suitable for blind tillage, giving a head start for the young crops against weed growth. Further weed control can be done using finger weeders, torsion weeders, rolling cultivations etc. that allow close tilling to the plant rows. Any smaller operations can utilize wheel hoes, stirrup hoes etc. for the same effect.

SOIL SOLARIZATION

The opening up of soil through tilling is followed by decontamination activity by soil solarization. It is basically a hydrothermal disinfection method. Soil made moist is covered up with transparent polythene sheets and exposed to direct sunlight when it is at its peak temperature for the year. Soil solarization is among the initial line of defense against soil borne pathogens like *Pythium* sp., *Phytophthora* sp., nematodes like *Heterodera* sp., *Xiphinema* sp., common weeds like *Cyperus rotundus*, *Cynodon dactylon* etc. After tilling, the soil is made clear of any debris and leveled for solarization. The beds should be then raised, as per crop requirements. After application of organic matter to the bed, it should be irrigated at the rate of 5 liters per m². The bed is then covered with 100-150 gauge transparent polythene sheet. Care should be taken to ensure that all corners and sides are properly sealed with the sheet. This will prevent any moisture loss or temperature fluctuation in the bed. After 20-30 days, the sheet can be removed and production practices started.

MULCHING

Mulching is the process of covering the topsoil with plant materials such as leaves, grass and crop residues. A mulch cover enhances the activity of soil organisms such as earthworms. They help to create a soil structure with plenty of smaller and larger pores through which rainwater can infiltrate easily into the soil, thus reducing surface runoff. Generally, cover crops, grass, weeds, crop restudies, pruning material from trees and hedges, wastes from agricultural processing or forestry are used as mulching materials. Plastic mulch, as long as it is removed at the end of growing or harvest season, is permitted in certified organic production. Mulching should be applied to before or at the onset of the rainy season, as far as possible. This is the period when soil is most vulnerable. If the mulch is applied prior to sowing or planting, the mulch layer should not be too thick in order to allow seedlings to penetrate it.

IRRIGATION SYSTEMS

Easy availability of water is essential for crop production. However, scarcity in its availability has led to development of techniques like drip irrigation and other water conservation techniques. Drip irrigation not only support reduced water usage, but also helps in weed management through selective application. The harvesting of water is followed especially in hilly terrains and also during the rainy season. The field preparation procedures like tillage, standing beds, crop density, crop cover etc. all influence the evaporation rate of water from the soil. The planning of an irrigation system should be done after consideration of all these factors and future requirements. Any system selected should not hamper the soil structure of the field, nor lead to excessing leaching of nutrients from the soil.

INTERCROPPING AND COMPANION PLANTING

Inter-planting of two or more mutually beneficial crops in close proximity is once way to increase biodiversity. On a large scale, this is called 'Intercropping'. The planning is done in such a way as to occupy different ecological levels. Along with preventing any competition, this will also help in controlling weeds in the field. It involves alternating rows or a number of rows of compatible field crops, like soybeans and corn. It also applies to sowing of forage crops like alfalfa, brome grass etc. When inter-planting is done on a smaller scale, it is called 'Companion planting'. For example, a mix of fruiting trees such as avocado, carambola, mango, jackfruit, with a density of 150 trees per hectare enhance cocoa production. Additionally, sapota and para-rubber trees (*Hevea brasiliensis*) can be interspersed.

GREEN MANURING

Green manure incorporates the idea of using specific live plants as nutritional replenishment for the soil. The green manure plants are selected based on the chemical content available to soil on their decomposition, their nature of growth and lifecycle. Details on few of the common green manure crops and their yield are given below.

- ✓ Sun hemp (*Crotalaria juncea*): This can be incorporated into the field 10 weeks after sowing of main crop. It cannot be used in water logging conditions. The seed rate is 25-35 kg/ha. Quantity of nitrogen fixed by the crop is 75-80 kg/ha. The green matter yield is 15-20 t/ha.
- ✓ Daincha (*Sesbania aculeate* and *S. rostrata*)
    - o *Sesbania aculeate*: This is a quick growing succulent green manure crop, incorporated at 8-10 weeks after sowing. It can withstand water logging,

drought, and salinity. Seed rate recommended is 20-25kg/ha. Green matter yield is 10-20 t/ha.
- *Sesbania rostrata*: This crop has nodules on the stem and root. It thrives under water logged conditions. The seed rate is 30-40 kg/ha. Green matter yield is 15-20 t/ha.

✓ Wild Indigo or kolingi (*Tephrosia purpurea*): This is suitable for light soil, resists drought but not water stagnation. The seeds have a waxy, impermeable hard seed coat and do not germinate fast. In order to induce fast germination, seeds have to undergo treatment with sand or hot water steeping for two to three minutes. The seed rate is 20-25 kg/ha. The green matter yield is 8-10 t/ha. Consecutive sowing for 2-3 seasons will give rise to volunteer plants and no further sowing is required.

✓ Indigo/ Bengal indigo (*Indigofera tinctoria*): It shows resistance to drought and is partial to clayey soil. Better yield is obtained when two irrigations are given. The seed rate is 20 kg/ha. Green matter production is 8-10t/ha.

GREEN LEAF MANURING

Green leaf manuring is an extension of green manuring, wherein shrubs and trees provide the sustenance to soil. Shrubs and trees are grown on field bunds or in other areas, topping branches from time to time for green manure. Such branches can be used as mulch, particularly in fruit orchards. The mulch conserves soil moisture in initial stages and consequently turns into organic manure. When the volume of dried leaves (litter) is high, it can be used directly as mulch instead of as manure. Selected shrubs and trees used for green manuring are given below.

- ✓ Gliricidia (*Gliricidia maculate*): This shrub is kept low by pruning or lopping at a height of 2-3 m. The pruning can be scheduled for twice of thrice a year. Each plant gives five to ten kg of green leaves annually.
- ✓ Subabul (*Leucaena leucocephala*): This is a branched shrub. It fixes about 500-600 kg N/ha per year.
- ✓ Cassia (*Cassia auriculata*): Propagated by seeds. During flowering, the tree is topped and loppings used for green leaf manuring.

CROP ROTATION

If the same crop is grown for several consecutive years on the same land, yields will normally decline and weed infestation, pest and disease incidences will increase. In order to prevent this, crop rotation is often recommended. Green manure crops should also be including in the cropping schedule for rotation. High demand crops should be followed by legume crops to replenish the soil. This break in heavy absorption of nutrients will help to extend use of fields. Some of the important cropping systems identified successful after field experimentation by the Network Project on Organic Farming (NPOF) of ICAR (Indian Council of Agricultural Research) are given below:

1. Rice-Wheat/mustard/lentil/potato

2. Soybean-Berseem/ Chickpea/Mustard
3. Tomato/Cabbage-Cauliflower-pea and maize-garlic
4. Maize-Cotton, Chili-Onion and brinjal-Sunflower

MANURING AND COMPOSTING

Certified organic producers have strict guidelines to follow in handling and applying manures. The National Organic Program (NOP) of U.S. regulations require raw animal manure incorporation into soil not less than 120 days prior to harvest of the product, whose edible portion has direct contact with soil surface or soil particles. It should be incorporated not less than 90 days prior to harvest in case of products whose edible portion does not have direct contact with the soil surface or soil particles. One of the best means of handling manures is composting.

Composting is a biological process involving the decomposition of organic matter by aerobic and anaerobic microorganisms. Vegetable and animal residues are generally used to prepare compost. Sewage sludge is prohibited from use in certified organic production. The process of composting involves preparation of a trench of suitable size (4-6 m length, 2-3 m breadth and 1-1.5 m depth). The accumulated residues are spread along the trench to 30 cm thickness, forming the first layer. This layer is well moistened by sprinkling cow dung slurry and water over it. A second layer of same thickness is spread, followed by moistening. This process is repeated until the heap reaches a height of 45-60 cm above ground level. The top is then covered with a thin layer of soil. Three months after this, the heap is opened and taken out, formed into conical shape and re-moistened with water. It is again covered up and left for the next 2 months. After that, the manure is ready for application to field.

VERMICOMPOSTING

Vermi-technology is a process by which all types of biodegradable wastes such as farm wastes, kitchen wastes, market wastes, bio wastes of agro-based industries etc. are converted to nutrient rich vermicompost by using earthworms as biological agents.

- ✓ <u>Selection of Species</u>: For vermicompost preparation, surface dwelling earthworms are to be used. The epigeic species have been found to be useful for compost making and most commonly used are Red worms (*Eisenia foetida*), Composting worm (*Peronyx excavatus*) and the African earthworm (*Eudrillus eugenial*). These species are fast breeders and feed actively on organic matter high in nitrogen.

- ✓ <u>Base material requirement:</u> Crop residues, tree leaves, animal dung, agricultural wastes like sugarcane trash, saw dust, coir waste, paddy husk, cattle dung, effluent slurry from bio-gas plant, excreta of sheep, horse, pig, poultry droppings (small quantity), vegetables waste etc. could be used as basic materials for vermiculture.

- ✓ <u>Containers for culturing:</u> Earthworm culture requires protection from shade and rain. Brick lined pits, plastic tubs, wooden boxes, earthen pots, or even heap of organic matter on surface of soil, can be used for culturing. The size of the container should be 1 m x 1 m x 0.3 m. In case of pit, heap method, dimensions may vary but depth should not exceed 45 cm.

✓ <u>Preparation of bed</u>: The preparation of compost and storage is summarized below:

- Step 1: Selection of container/ digging or pit for culturing (as per above specified dimensions)
- Step 2: Raise a bed of 10cm height using any of the base materials selected (coir waste, paddy husk etc.). Cover it with a layer of soil and sprinkle water on top, will it attains 40-45% moisture level.
- Step 3: Make a homogenous mixture of organic waste, cattle dung and water. Keep this mixture for 2 weeks. Turn the material 2-3 times at 4-5 days interval and then transfer to the bedding layer prepared (Step 2).
- Step 4: Introduce the worms or cocoons to the bed at the rate of 2000 worms for 400 kg of feed mix (Step 3). If culturing is being done for the first time, it is advisable to use worms. The feed mix is to be spread uniformly on the bed, adding 5-10% neem cake into it.
- Step 5: Cover the bed with gunny cloth. Maintain moisture level of the bed by sprinkling water periodically over the cloth. The worms will convert the feed mix into vermicasting in 60 days.

- Protect the worms against red ants, centipedes, toads, cats and other predators during this time.
- Step 6: Take out the vermin compost and heap in sunlight over a plastic sheet. Keep it undisturbed for 1-2 hours. Remove the compost at the top, and collect the worms settled on the bottom for next batch of vermicomposting. Screen method can also be used to harvest the worms. A box with a screen bottom can be used to separate the worms from the compost.
- Step 7: The harvested vermicompost should be stored in a dark, cool place with minimum 40% moisture. It is preferred to store the compost openly and only packing it during point of sale. Vermicompost can be stored for a period of 1 year without any quality loss.

COIRPITH COMPOSTING

Coir pith is a waste material from the coir industry. Its wide C: N ratio and rich lignin content does not allow its natural composting. Mushroom *Pleurotus sp.* has the capacity to degrade part of the cellulose and lignin present. Coirpith (1 ton), poultry manure (10kg), mushroom *Pluerotus* spawn (1.5 kg) is required for preparation. A shaded area of 5 m x 3 m is to be selected, cleared of weeds and leveled. 100kg of coir pith has to be spread uniformly. Spread 300 g of *Pleurotus* spawn on top of this and cover with a layer of coir pith (100kg). Spread 2 kg of poultry manure uniformly on surface of the second layer. Repeat this sandwiching process up to 1 m height. Sprinkle water to maintain moisture. After leaving the heap undisturbed for 2 months it can be used.

PANCHAKAVYA

It is an organic product, which promotes growth and immunity in the plant system. It consists of nine products namely cow dung, cow urine, milk, curd, jiggery, ghee, banana, tender coconut and water. The process involves the following steps:

- ✓ Step 1: Mix Cow dung (7kg) and Cow ghee (1kg) thoroughly in morning and evening hours and keep for 3 days.
- ✓ Step 2: Mix cow urine and water to Step 1 after 3 days and keep for 15 days. Mixing should be done regularly in morning and evening hours.
- ✓ Step 3: Mix Cow milk (3 lt), Cow curd (2 lt), Tender coconut water (3 lt), Jaggery (3 kg) and well ripened banana (poovan) 12 nos, to the above mix and leave. The panchakavya will be ready within a month.

Application

3 liter of Panchakavya to every 100 liters of water is ideal for spray system for all crops. In case of drip or flow irrigation, the solution of Panchakavya should be mixed with 50 liters/ha of irrigation water.

Seed/seedling treatment

3% solution of Panchakavya can be used to soak the seeds of dip the seedling before planting. Soaking time of 20 minutes is sufficient. Rhizomes of turmeric, ginger, setts of sugarcane can be soaked for 30 minutes before planting.

The application of Panchakavya induces drought hardiness in plants. The formation of a thin oily film on the leaves and stems reduces evaporation of water. Increase in leaf size, development of denser canopy, production of side shoots, root densing etc. are other effects seen from Panchakavya usage.

DASAKAVYA

Dasakavya is the combination of Panchakavya and plant extracts. The plants used are: *Azadirachtaindica, Calotropis*sp.*, Tephrosia purpurea, Vitex negundo, Datura metel, Jatropha curcas, Adathoda vasica, Pongamia pinnata*. The plant extracts are prepared by separately soaking the foliage in cow urine in 1: 1 ration for ten days. The filtered extracts of all the plants are then added at the rate of 1 liter each to 5 liters of the Panchakavya solution. The mixture is kept for 25 days and stirred well, after which it is ready to use.

BEEJAMRUT

Beejamrut is generally used for seed treatment. It is prepared by mixing together 5 kg of cow dung, 5 liters of cow urine, 1 liter of cow milk, 250 g of lime and 100 liters of water. The formulation is kept overnight and can be used from the next morning. The seeds sprinkled with Beejamrut should be shade dried before sowing.

JEEVAMRUT

Take 100 liters of water, 10 kg cow dung, 10 liters of cow urine and mix in a barrel with a wooden stick. Once thoroughly mixed, add 2 kg jaggery and 2 kg flour to it. After mixing, the solution must be allowed to ferment for 7 days. In order to facilitate complete fermentation, the vessel should be shaken at least 3 times daily.

BIOFERTILIZERS

Biofertilizers can be defined as preparations containing living cells or latent cells of efficient strains of microorganisms that help crop plants' uptake of nutrients by their interactions in the rhizosphere when applied through seed or soil. They accelerate certain processes in the soil which augment the extent of availability of nutrients in a form easily assimilated by plants. The use of biofertilizers is one of the most important components of integrated nutrient management (INM), with its cost effectiveness, and renewable nature.

-There are different groups of biofertilizers available:

1. <u>N2 fixing biofertilizers</u>: *Azotobacter, Rhizobium, Azospirillum, Clostridium*
2. <u>P solubilizing biofertilizers</u>:*Penicillium.* sp., *Aspergillus awamori, Bacillus megaterium* var. *phosphaticum*
3. <u>P mobilizing biofertilizers</u>: *Rhizoctonia solani, Pezizellaericae, Laccaria*sp.,*Glomus* sp.
4. <u>Biofertilizer for micro nutrients</u>: *Bacillus* sp.

5. <u>Plant growth promoting rhizobacteria:</u> *Pseudomonas fluorescens*
6. -Rhizobium *(Bradyrhizobium* and *Azorhizobium)*: It is a symbiotic nitrogen fixer, supporting fixation of nitrogen in leguminous plants. It induces better root nodulation and stem nodulation in inoculated plants, thereby reducing the nitrogen fertilizer requirement for pulses, oil seeds and legume green manures. Method of application is seed treatment.
7. Azotobacter: It is a free living nitrogen fixer. Plants inoculated with azotobacter reap benefits of increase in plant biomass, nitrogen uptake, development and branching of roots etc. Method of application are seed treatment, seedling dip and direct application to soil through organic manure.
8. Azospirillum: It is an associative micro-aerophilic nitrogen fixer. It colonizes in areas of low oxygen tension in root mass. It induces plant roots to secrete mucilage for aeration in such environment. It is thereby suitable for both upland and wetland conditions and is available as carrier based inoculums. Under ideal conditions, it fixes around 20-25 kg per ha of nitrogen. Treatment with *Azospirillum* also induces better root formation in inoculated plants.

Method of application is given below:

Seed treatment: 500 g culture for treating of 5-10 kg seeds. Moisten the seeds by sprinkling water or rice-gruel water, Take 500 g culture in a plastic tray/basin, add moistened seeds, mix well and dry in shade for 30 minutes. This can be sown immediately.

Seedling root dip: Slurry of the culture is prepared by mixing 500 g culture with 750 ml of water and the roots are dipped in the slurry for 15-20 minutes before transplanting.

Soil application: Mix the culture with FYM or compost in the ratio 1:25 and apply directly to soil.

Inoculation for paddy: Mix 2 kg of culture in 60 liters of water and soak seeds required for 1 ha (60 kg) for 24 hours before sowing. At the time of transplanting, dip the roots of seedlings for 15-20 minutes in the culture slurry prepared by mixing 2 kg inoculum with 40 liters of water, Thus slurry can be used for treating seedlings required for 1 ha. Another 2 kg culture may be applied to the field along with FYM or compost

9. Phosphate solubilizing bacteria *(PSB):* Indian soils are low-medium in available phosphate. Bacteria like *Bacillus megatherium, Bacillus polymyxa, Pseudomonas strata, Pseudomonas rathonis* help in better absorption of phosphate from the soil. Inoculation of the same reduces need for phosphate application by 50%. The method of application is seed treatment and direct application through organic manure.
10. Potash mobilizing bacteria (*Frateuria aurentia*): K-Bacteria helps release gorged potassium between layers of clay.

Potash mobilization at the rate of 40-60 kg per ha can be achieved by application of this gram negative rod type bacteria.

11. Azolla and blue-green algae: Blue-green Algae, in symbiotic relationship with water fern belonging to *Azolla* sp., fix atmospheric nitrogen in the soil. *Nostoc* and *Anabaena* are the two popular species of blue-green algae. It can contribute over 100 kg of nitrogen per hectare per crop like paddy under low land conditions. Blue-green algae available as a carrier based inoculum and is directly broadcasted to rice fields at 10kg/ha one week before transplanting of seedlings. Azolla can be cultivated in shallow cement tanks or pits lined with polythene sheets. Pits or tanks of convenient length and breadth and 15 cm depth are made. Soil has to be spread with uniform thickness at the bottom of the tank at the rate of 7 kg per m^2. Fresh cow dung at the rate of 2.5 kg per m^2 has to be made into slurry and poured uniformly on the soil in the tank. Rajphos or Mussoriphos at the rate of 15 g per m^2 has to be given along with cow dung slurry. Water has to be added to the tank to a depth of 8 cm. Healthy azolla at the rate of 250 to 500 g per m^2 has to be spread uniformly in the tank. Azolla starts multiplying after a period of one week. For large scale field application, azolla nursery can be raised in the field itself. Field should be ploughed thoroughly and leveled uniformly. The field is divided into one cent plots (20 x 2 m^2) by taking bunds and channels.

Water is maintained at 10 cm level. Fresh cow dung slurry prepared by mixing 8 kg of cow dung in 20 liters of water is sprinkled over the plot. Fresh azolla inoculum is spread @ 8 kg per plot. Rock phosphate can be applied @ 100 g in three split doses at 4 days interval. 15 days after inoculation, Azolla is harvested and applied to main field. Fresh azolla can be applied at the rate of 10t/ha before transplanting of rice seedlings.

PEST AND DISEASE MANAGEMENT

Pest management (IPM) and disease management (IDM) involves use of mechanical, cultural, biological methods to prevent or control pest and disease attack on the organic crops. Since application of preventive chemicals is prohibited, it is essential to take other steps to control any incidence. Crop monitoring for incidences and taking up of preventive measures cover a large portion of IPM and IDM activities. The correct prognosis of type of pest/disease incidence is also essential for further management process. It is often seen that proper pest management brings down disease incidence. This is attributed to the removal of pests transmitting/transferring disease causing pathogen in the field, for example spread of virus by whiteflies and leaf hoppers. The initial selection of a healthy and appropriate seed stock variety is one such preventive step. The other activities included in IPM and IDM are given below:

1. -Tillage
2. -Soil Solarization

3. -Seed treatment
4. -Crop rotation
5. -Trap crops
6. -Weeding
7. -Mulching and mowing
8. -Fertilization management
9. -Irrigation control
10. -Biocontrol agents
11. -Biological preparations
12. -Permitted fungicides like Bordeaux mixture (1%)
13. -Fumigation
14. -Mechanical controls like traps, barriers

Trap Crops

Trap crops are lures to wean pests away from the main crop, in case of a pest attack. The pests are either prevented from contacting the main crop or are limited to pockets across the field. The selection of trap crops is based on past preferences for certain crops by pests. A most common example is the use of African marigold as trap crop for nematode control. The benefit of using trap crops is that the main crop gets limited to nil pest damage. This could be related to expected yield maintenance. Further, the concentration of pests to pockets may also attract its natural enemies to the site, enhancing the natural bio control mechanism. However, trap cropping must be done only when there is increasing chances of attack by a high pest population. Otherwise, it might lead to formation of pest nurseries and create further problems.

Fumigation

Fumigation is an essential sanitization process for disease control. About 200g of Embeliaribes or Acorus calamus (sweet flag) should be powdered well and placed in a wide mouthed pot along with burning charcoal. This should be carried across the field in opposite wind direction. Sweet flag rhizomes extract should be sprayed 7 days after fumigation. This will help in controlling bacterial and fungal diseases to an extent.

Bio-control agents

1. Arbuscular Mycorrhizal Fungi (AMF) : Inoculation with AMF at time of planting improves the growth and tolerance of crops to root pathogens, especially *Phytophthora, Pythium, Rhizoctonia* and root nematodes of black pepper, cardamom, ginger, turmeric, rice, transplanted vegetables.

2. Trichoderma: Biocontrol of soil borne plant pathogens involves introduction of antagonistic microorganisms in the soil. Trichoderma is effective against quick wilt of pepper (*T.viride* T6, *T.longibrachiatum* T2), rhizomes rot of cardamom (*T. longibrachiatum* T2, *T.virens* T9) and ginger (*T.viride* T10). Neem cake-cow dung mixture is used as food base for *Trichoderma* sp.

3. Fluorescent pseudomonas: They are a group of bacteria very effective against disease incited by species of *Phytophthora, Pythium, Rhizoctonia, Fusarium, Colletotrichum, Ralstonia* and *Xanthomonas* in various crop plants.

4. Bacillus thuringiensis (Bt): Spraying of *Bacillus thuringiensis* (Bt), a pathogen capable of stacking several *Lepedoptera* sp. pests have been successfully adopted to control caterpillars, beetles in vegetables and other agricultural crops, and for mosquito and black fly control.
5. Nematodes: Nematodes such as *Steinernema carpocapsae* control soil insects such as cutworms (*Agrotis* sp.) in vegetables. *Phasmarhabditis hermaphrodita*, a microscopic nematode kills slugs, giving up to 6 weeks of protection on application.
6. Natural Predators: Use of natural enemies should be limited to cases of large scale pest invasion. The selection of predators is linked to both their huge consumption rate and non-conversion into pests for main crop. Ladybugs and their larvae prey on aphids, mites, scale insects and small caterpillars. The larvae of some hoverfly species feed on greenfly, fruit tree spider mites and small caterpillars. *Typhlodromips swirskii* is useful against spider mites, thrips and white flies, while *Phytoseiulus persimilis* against spider mites. Other useful natural enemies include pirate bugs, rove and ground beetles, spiders, syrphid flies, predatory mites, frogs, toads, birds, slow-worms etc. Field mice and rats can be controlled using cats and rat terriers.
7. Parasitoid insects: Parasitoids lay eggs on insect host body, later suing same for nourishment of larvae. In this manner, the host is ultimately killed. Most of the parasitoids are either wasps or flies with very specific host requirements.

Chalcid wasps parasitize eggs of greenfly, whitefly, cabbage caterpillars, scale insects and strawberry Tortrix moth (*Acleris comariana*). Tachinid flies parasitize on caterpillars, adult and larval beetles, true bugs etc. Another example is the *Encarsia formosa*, a predatory chalcid wasp which parasitizes on whitefly.

Biological pesticides
1. Neem as pest repellent: Pound neem leaves/ neem cake/ neem kernels and place in a pot. Add twice the volume of water and tie the mouth of pot with cloth. Leave it for 3 days. Later, place the pots on four corners of the field. In the evening open the pots. The foul smell from the neem product will prevent entry of pests into field.
2. Ginger, Garlic and Chili extract: 1 kg of garlic should be immersed in 100 ml of kerosene and kept overnight. Next day. The outer skin should be removed and made into a paste. In another vessel, ½ kg of chili should be mixed with 50 ml water and made into a paste. Likewise, ½ kg of ginger should be made into a paste. All the three mixtures should be mixed together with 100 liters of water and 50 g of soap solution as emulsifier. This mixture should be stirred well and filtered before spraying. The above quantity is for an acre. Allicin present in garlic serves as a repellent and capsaicin in chili serves as a pesticide.
3. Tobacco and chili extract: Take the remains left after harvesting tobacco leaves and pound into powder. Mix 2 kg chili powder with 3-4 kg tobacco powder and add 5 kg

of sand to it. Dust it over the plants early morning. The above quantity is for an acre.

4. Some of the components allowed for disease and pest control in organic farming include sulphur (fungal diseases, spider mites), copper (fungal diseases), sulphuric acidic argillaceous earth (fungal diseases), ashes (soil borne diseases), slaked lime (soil borne diseases), clay (fungal diseases), baking soda (fungal diseases), soft soap solutions (aphids and other sucking insects), plant ashes (leaf miners, stem borers). Quick lime, chloride of soda/lime, permanganate of potash, sea weeds, natural acids etc are among the many not permitted for pest and disease control in organic farming.

5. Neem oil: 25 to 30 ml of neem oil should be mixed with soap water to form an emulsion. This emulsion spray will help against fungal diseases like Downey mildew. It is also effective against beetles, plant hoppers, caterpillars etc. Seed protection during storage is also one of its objectives. For every 500 ml of seeds, mix 5-10 ml of oil and store in air tight containers.

Mechanical control

1. Light trap: Light traps can be used to monitor and trap the adult moths thereby reducing the population. Some of the most common light traps that could be used are hurricane lamp, trap with electric bulb etc. The adult moths have an inherent capacity to get attracted to the light. It should be set up in field after 5:30 pm. A large plate or vessel of

kerosene mixed with water is kept near the trap. The attracted moths fall into this and die.

2. Yellow sticky trap: Castor oil smeared yellow color empty tins or plates are kept in the field. White flies get trapped on these. These can be wiped out every day and castor oil applied again.

3. Pheromone traps: Pheromones can be used for mass trapping of pests, for monitoring the insect population, to disrupt mating, so as to control further spread of the pests. The attractants are used in various trapping mechanisms like sticky, wire, mesh, pan and water traps. The traps should be set 2 feet above the crop canopy. The high cost of pheromones, however prevents its regular use by farmers.

4. Bird traps: Attracting birds
   - ✓ Bird perches: Install "T" shaped bird perches which are long dried twigs at the rate of 15-20 per acre. These attract birds for resting and the resting birds devour the larvae in the field.
   - ✓ Bird attractant: 1 kg of rice and 50 g of turmeric powder is required per acre. The rice is cooked and excess water is filtered. This is mixed with turmeric powder. Small lumps of yellow colored rice is taken in small vessels and placed in 8-10 places in the field. This is kept during early morning and afternoon. This acts as bait for attracting birds. While the birds feed on the rice, they also feed on the semilooper larvae

prevalent in the field. This procedure is repeated till the crop attains the flowering stage thereby reducing the pest attack.

5. Hand picking method: This is feasible only if the field size is small. Pour a small amount of kerosene in a polythene bag and pick up the larvae during evening hours and put it in the bag. This should be done when the pest numbers are low. Wild grasses and weeds should be removed from the field bunds and field, since these are the favorite egg laying sports of pests.

6. Rat control: To control rats, pieces of papaya are spread near the bunds of the field. Papaya has a chemical substance which causes tissue damage in the mouth of the rats feeding on it. For an acre, 3 fruits are required.

WEED MANAGEMENT

Weeds are plants that grow in places where they are not wanted or in unwanted periods of the cropping system. A basic working principle in organic farming is to prevent problems, rather than to cure them. Good weed management in organic farming includes creating conditions which hinder weeds from growing at the wrong time and place. Since the use of synthetic herbicides is prohibited, it is essential to follow back on these cultural, mechanical methods of preventive weed management. Some of the basic preventive measures are maintaining green cover, mulching, crop rotation of high demand crops with less demanding crops like legumes, balanced fertilization, soil cultivation, soil solarization, irrigation etc. A living green cover will prevent weed growth and spread through competition for nutrients, water and space. Mechanical control of weeds can be done by manual removal, flame weeding etc. Proper heating of compost while preparation also eliminated weed seeds if any from reaching the field. The time of seed sowing and spacing of sowing also influence growth of weeds. Tillage of the field after harvest and during the summer time is essential to expose the layers of soil to high temperature, thus eliminating any weeds, pathogens, pests etc. Only a limited number of herbicides are allowed in organic cultivation. These include acetic acid (vinegar), citric acid, solutions of sodium nitrate and corn gluten (pre-emergent). Weeds that emerge before the main crop can be killed using the contact herbicides like acetic acid. Pre-emergent herbicides are applied to soil to suppress any germination of weeds. Weed management is essentially crucial for organic farming because of increase in chance of contamination. Care should be taken to prevent any weed spread from nearby fields, especially if they follow the conventional farming system. Also, attention should be given while harvesting and processing of crops, especially grain crops for weed seed contamination.

ORGANIC FOOD PROCESSING AND HANDLING

Harvesting of organic produce is done with the concept that the quality at time of harvest is the best. In order to maintain quality of produce harvested, it is recommended to harvest during the coolest time of the day and to keep the produce in shade. Handling of the organic produce must be done in an optimal manner to maintain the quality of the product and also minimize development of pests and diseases. The adoption of Good Agricultural Practices (GAP) measures will help in reducing chances of contamination and other hazards for organic produce. Standard GAP procedure for sanitation and water disinfection needs to be followed before processing. Processing and handling of organic produce must be done separately from non-organic produce. The processing room should be sanitized and free from any form of contaminants. Recognizing the exceptions to processing ingredients in organic food processing is essential. The flavoring extracts used must be organic in nature and from natural sources. Maintenance of a controlled atmosphere, cooling, freezing, drying, humidity regulation and storage at ambient temperature process are some of the conditions permitted for organic foods. Ethylene gas is permitted for use in ripening process. Water and salt are permitted for use in organic food processing. Usage of minerals, vitamins and similar isolated ingredients is not allowed (exceptions when need proved to certification agency). There is a limit to the processing practices permitted. Permitted processes are mechanical and physical, biological, smoking, extraction, precipitation and filtration. Extraction should only be done with water, ethanol, plant and animal oils, vinegar, carbon dioxide, nitrogen or carboxylic acids. Irradiation is strictly prohibited.

ORGANIC FOOD PACKAGING AND LABELING

The marking and labeling of the produce must be in accordance with the provisions of the National Program for Organic Production. Packing and labeling protocols need to be followed for organic food distribution in the market, especially for export. The packaging materials used for organic produce should be eco-friendly. Any unnecessary packing is to be avoided. The ink used for marking on package should be of quality, which may not contaminate the produce. Contamination of food produce due to packaging material used should be avoided at all costs. The use of PVC materials is prohibited and laminates and aluminum are to be avoided. The claimed food product should be labeled as per certification agency specifications as 'Organic' with logo 'India Organic' for export. In addition to the grade designation mark, the information on season of harvest, expiry date, place of packaging, name of packer etc. should also be made available on each package. Water and salt should not be included in the percentage calculations of organic ingredients

ORGANCI FOOD STORAGE AND TRANSPORT

The organic food produce during commute and storage should be kept away from contact with any non-organic produce. A separate storage room should be made available to separate the organic-non organic produce. Care should be taken to prevent organic produce cargo exposure to any materials or substances not permitted in organic farming.

www.ingramcontent.com/pod-product-compliance
Lightning Source LLC
Chambersburg PA
CBHW070729180526
45167CB00004B/1674